GENERAL MOTORS TYPE 5: CLASS 66 LOCOMOTIVES

Ross Taylor

AMBERLEY

First published 2015

Amberley Publishing
The Hill, Stroud
Gloucestershire, GL5 4EP

www.amberley-books.com

Copyright © Ross Taylor, 2015

The right of Ross Taylor to be identified as
the Author of this work has been asserted in
accordance with the Copyrights, Designs and
Patents Act 1988.

ISBN 978 1 4456 4850 7 (print)
ISBN 978 1 4456 4851 4 (ebook)

British Library Cataloguing in Publication Data.
A catalogue record for this book is available from
the British Library.

Typesetting by Amberley Publishing.
Printed in the UK.

Contents

Introduction

My intention for this publication is to provide a detailed pictorial history of General Motors Type 5 Class 66 diesel electric locomotives. Following the privatisation of the UK's rail network, the American company Wisconsin Central bought most of the British Rail freight sectors. These were Rail Express Systems, Transrail, Loadhaul, Mainline and, later, Railfreight Distribution. This company then became known as EWS, which stands for English, Welsh & Scottish Railway.

EWS inherited an elderly fleet of various types of locomotives, many of which were unreliable and costly to maintain. This led to the decision to order a brand-new fleet of diesel locomotives, which was fulfilled by General Motors of the USA. The order, which was placed in May 1996, was for 250 locomotives to be known as the Class 61. Thirty of these locos were to be fitted with ETH (Electric Train Heat) and high-geared traction motors and would be classified as the Class 66. However, by October 1996 this order had changed to 250 identical locomotives, reclassified from Class 61 to Class 66, to be numbered 66001–250. The thirty high-geared locomotives that were going to be the original Class 66 formed a separate order to become the Class 67.

The Class 66 design was based mechanically on General Motors (GM) SD70 MAC American diesel locomotives but fitted with the latest electronics from the SD80 MAC. Bodywork was based on the earlier Class 59. The first Class 66 to arrive in the UK was 66001 on 18 April 1998, which touched down on UK soil at ABP Immingham Dock at 8.53 a.m. The locomotive was fitted with prototype bogies, which were replaced by production ones later in the year. The last EWS Class 66 to arrive was 66250 on 21 June 2000.

Following the successful operation of the class by EWS, rival company Freightliner ordered five, which arrived in the country in July and August 1999. These locomotives arrived at the same time as a small batch of EWS locos and, as with all the EWS 66s, they were transported by the ship *Fairload*, operated by Jumbo Shipping. This ship could hold up to eleven locomotives at one time.

Freightliner has gone on to become the second largest operator of the Class 66, having ordered 131 locomotives from new. However, one extra was ordered to replace 66521, which was involved in a serious collision at Great Heck in north Yorkshire on the East Coast Main Line on 28 February 2001. 66521 arrived in the UK on 12 December 2000 and only lasted seventy-eight days, which was one of the shortest working lives of any rolling stock in the country. It was finally scrapped at C. F. Booth of Rotherham in June 2006.

As a newcomer to the UK's rail network, GB Railways announced an order of seven standard Class 66/7 locomotives in early 2000. These arrived as 66701–707 in March 2001, and over the following years, leading up to 2008, further Class 66/7s were ordered in line with the expansion of the company. The locomotives arrived in several small batches, finishing in April 2008 with 66732.

In 2011, GBRF (the newly renamed GB Railways) took a further thirteen second-hand locomotives from operators Freightliner, Colas Rail and Direct Rail Services (DRS). These were renumbered into the 66/7 series as 66733–746 and will be covered in the chapters on their respective operators. In addition, GBRF purchased three unused locomotives from Crossrail AG in the Netherlands. These arrived in the UK in December 2012, becoming 66747–749. Two further second-hand locomotives were also purchased from Europe, these became 66750 and 66751. All five of the European Class 66s have some differences compared to the British locomotives and initially only Tyne Dock and Liverpool drivers were trained on them.

With the impending discontinuation of the Class 66 design due to tightening of European emission regulations (which the class cannot meet), GBRF took the decision to order a further twenty brand-new locomotives. These new locos were delivered in the latter half of 2014 numbered 66752–772. However, due to a loophole in the law that applies to the power units registered in the UK before the deadline of 1 January 2015, GBRF have been able to order a further five locomotives.

Three will have new power units, which were brought to the country for registration, and two will have second-hand units from scrapped locomotives. These locomotives are to be built, when required, at a later date and will be numbered 66773–779.

Direct Rail Services (DRS), based in the Cumbrian city of Carlisle, decided to follow other operators in ordering a fleet of their own Class 66/4. They started with an order for ten locomotives, which were numbered 66401–410, and arrived in two batches of five in October and November 2003. After the success of these, DRS decided to order a further ten of the low emission variant in summer 2004. These also arrived in two batches in May and October 2006 and were numbered 66411–420. After the delivery of these a third order was placed, again for ten locomotives, and these were delivered in September 2007, numbered 66421–430. Before these arrived a further four locomotives were ordered which arrived in November 2008 and numbered 66431–434.

Following deliveries of the later batches for DRS, it was decided to dispose of earlier locomotives to other companies as follows: 66406–409 were sent to Advenza while 66410 went off lease and was stored. Nos 66401–405 went straight to GBRF between March and June 2011 and Freightliner took 66411–420 between September and November 2011. DRS now run nineteen Class 66s, including the five Class 66/3s formerly operated by Fastline Rail, who went into liquidation in March 2010. They are numbered 66301–305 and were new between June and November 2008.

Advenza Rail Freight was founded in 2001 and was part of Cotswold Rail. They took ex-DRS locomotives 66406–409 and renumbered them as 66841–844 in the summer of 2009. However, the company went into administration and the locomotives went on spot hire to other operators as follows. 66841 and 66842 went to Colas Rail, 66843, which never operated in Advenza service, and 66844 went to GBRF for a few months at the end of 2009 before moving to Colas Rail.

Colas Rail is a French rail freight company formerly known as Seco Rail. After purchasing three older locomotives of Class 47 (originally the Brush/Sulzer Type 4), the company wished to add newer locomotives to their fleet. The former Advenza Rail Nos 66841–844 were used until these transferred to GBRF in July 2011, when

Freightliner 66573–577 were overhauled, renumbered 66846–850 and taken on by Colas Rail. The Colas 66s can normally be found working timber trains around the north-west to Chirk and in the south-east of England.

In July 2007, infrastructure company Jarvis created a new operator for the British network: Fastline Rail. They ordered five new Class 66/3s that were numbered 66301–305. The company only survived until March 2010 when the whole Jarvis Group became bankrupt. As previously mentioned, these locomotives were subsequently leased to DRS.

In the UK the various operators operate the following numbers of locomotives (as of January 2015):

DB Schenker: 176
Freightliner: 118
GBRF: 71 (with five more to follow)
DRS: 19
Colas: 5

Both DB Schenker and Freightliner also have European operations and have transferred seventy-four and thirteen locos respectively to Europe. The Freightliner ones including former DRS 66411/412/417.

In addition, the Class 66 design has enjoyed considerable success with European operators ordering batches to slightly modified specifications, but these are outside of the scope of this book.

This, then, is the story of the Class 66 to date (April 2015), but the class has many more years of service yet to come in which there will doubtless be further developments.

Advenza

No. 66841 sits at Shipley awaiting its next turn, which will be the following day's 6V67 13.45 to Cardiff Tidal loaded scrap train. Advenza rail freight was very short lived. Photographed here on 11 October 2009.

Advenza Rail 66841 sits next to preserved BR blue grid No. 56006 at Barrow Hill. The occasion was the diesel gala over the weekend of 8/9 August 2009.

Colas Rail

Colas Rail No. 66843 is seen powering past Birkett Common, near Kirkby Stephen, on the Settle–Carlisle line while in charge of the 6J37 Carlisle Yard–Chirk Kronospan loaded timber train. The train is seen here on 14 April 2010. This Class 66, along with Nos 66841, 842, 844 and 845, have now gone to GBRF and are numbered 66742–746.

Seen here at Ribblehead Virtual Quarry is Colas Rail No. 66850, the locomotive was former Freightliner No. 66577. The train will work 6V37 to Chirk Kronospan later in the week. Seen here on 8 October 2011.

After being named *Wylam Dilly* on 2 August 2011 at Stanhope station on the Weardale Railway, Colas Rail No. 66849 tackles the steep gradient at Wilpshire in Lancashire while in charge of the 6J37 Carlisle Yard–Chirk Kronospan loaded timber train. Photographed here on 20 April 2015.

No. 66849 is seen here after Colas Rail successfully won the contract off DB Schenker for the 6M32/6E32 Lindsey Oil Refinery to Preston Docks (Ribble Rail) and return, and is pictured as a star guest to the Ribble Steam Railway's diesel gala. It is seen here on the Ribble Steam Railway at Preston Docks while working a short shuttle service down the branch on 7 March 2015.

No. 66849 is seen arriving into Carlisle Citadel station while in charge of the 6J37 Carlisle Yard–Chirk Kronospan loaded timber train. Seen here on 11 April 2015.

Arriving into Elford goods loop is No. 66848, in charge of the empty 4V30 Ratcliffe-on-Soar power station–Portbury coal terminal coal train. Seen here on 16 May 2014.

Taking a short break at Preston is Colas Rail's General Motors No. 66847, in charge of the diverted 6J37 Carlisle Yard–Chirk Kronospan loaded timber train. Photographed here on 7 July 2014.

No. 66849 *Wylam Dilly* passes through Clitheroe while in charge of the 6J37 Carlisle Yard–Chirk Kronospan loaded timber train. Seen here in the autumn light on 26 September 2014.

DB Schenker

DB Schenker No. 66129 passes Slindon with the 6M66 Southampton Docks–Halewood loaded car train. Seen here in the late afternoon sun on 12 June 2012.

No. 66198 slowly crosses over Fairwood Junction near Westbury while in charge of the 6M20 Whatley Quarry–Hayes and Harlington loaded stone train. Photographed here on 14 May 2014.

No. 66129 is seen powering past Burton Salmon while in charge of the 6M96 Milford West sidings–Tunstead empty stone train. Seen here on 21 June 2014.

EWS-liveried No. 66183 creeps round the corner at Long Preston as it approaches its booked stop at Hellifield goods loop while in charge of the 6E93 Hunterston–Ferrybridge loaded coal train. Photographed here on 23 July 2014.

No. 66120 shows that the class can provide some thrash as it powers past Burton Lane with the 6H26 Butterwell Colliery–Drax loaded coal train. Seen here on 21 July 2014.

Seen rounding the curve away from Hellifield is No. 66113 as it powers away from a crew change while in charge of the 4M00 Mossend Down Yard–Clitheroe Castle cement empties. Seen here on 23 July 2014.

No. 66075 is seen heading through Shrivenham while working the 6V47 Tilbury–Trostre steel train. Photography here will soon be ruined due to Network Rail installing overhead line wires in preparation for the new electric trains. Photographed here on 15 May 2014.

After arriving into the country in 1998, some seventeen years ago, No. 66009 creeps through Knottingley station while in charge of the 6E56 Tunstead–Drax loaded limestone train. Photographed here on 8 April 2014. The former National Power wagons are now being repainted in DB Schenker red livery.

Normally hauled by a DB Schenker Class 60 super tug is the 6M00, which runs from Humber to Kingsbury and is loaded to 3,000 tonnes. However, a shortage of Class 60s at Immingham resulted in No. 66164 being allocated to work the train seen here passing Melton Ross limeworks. Photographed here on 12 March 2014.

No. 66050 *EWS Energy* pulls away from its booked stop at Sudforth Lane while in charge of the 4A70 Drax power station–Milford sidings empty coal working. Seen here on 13 March 2014.

No. 66050 *EWS Energy* is seen powering past Burton Salmon while in charge of the 6H84 Hull–Drax power station loaded biomass train. Photographed here on 13 March 2014. The driver is seen giving a thumbs up for the photographers.

The newly painted DB Schenker No. 66118 passes through Doncaster train station on the fast lines while working the 4L45 Wakefield Europort–Felixstowe container train. Pictured here on 13 August 2013.

Double-headed Class 66s Nos 66098 and 66198 pass Settle Junction with the very early 6K05 Carlisle Yard–Crewe Basford Hall departmental service. The train is now run by Direct Rail Services (DRS). Photographed here on 1 April 2013.

No. 66071 is seen on the outskirts of Loughborough while in charge of the 6M23 Doncaster Virtual Quarry–Mountsorrel empty ballast boxes. Seen here on 18 August 2012. DB Schenker relocated these locomotives back to the UK for short-term hire when there was a shortage of locomotives due to preparation for the RHTT trains in the autumn.

The new DB Schenker red livery is seen adding some bright colour into the north Yorkshire countryside as No. 66101 passes Newsholme with the 4Moo Mossend Down Yard–Clitheroe Castle cement empties. The train was photographed here on 19 September 2012.

Now run by Euro Cargo Rail DB Schenker in France, No. 66228 is seen passing Brock with the 6K05 Carlisle North Yard–Crewe Basford Hall departmental service. The train was photographed here in the late evening sun on 25 September 2006.

No. 66227 has a rest as it sits in Hellifield goods loop awaiting a crew change. The train was the 6L65 Clapham (north Yorkshire)–Crewe Basford Hall engineering train, pictured here on 14 March 2010. The Class 66 has now left the country and runs in France for Euro Cargo Rail, leaving only eleven of the 662xx locos in Britain.

No. 66015 passes Newsholme in north Yorkshire with the Settle Junction–Crewe Basford Hall 6L62 engineering train. Seen here on 14 March 2010.

Now run by DB Schenker but still in EWS livery (as is the majority of the class) is Class 66 No. 66102, seen passing Red Bank near Newton-le-Willows with the daily 6K05 Carlisle North Yard–Crewe Basford Hall departmental service. The train was formed of the HOBC, which stands for Network Rail's High Output Ballast Cleaner. Pictured here on 10 August 2012.

No. 66081 powers past Sherburn-in-Elmet during a brief spell of sun while working the 13.40 6H26 Redcar BSC–Eggborough power station loaded coal train. Seen here on 8 April 2014.

No. 66012 rounds the corner at Kings Sutton while working the 4M52 Southampton Eastern Docks–Castle Bromwich Cartics train. Seen here on 15 May 2014.

No. 66132 is seen arriving into Westbury while working the 7C48 Appleford sidings–Westbury empty stone train. Seen here on 14 May 2014.

No. 66043 thrashes past Elford goods loop while running late with the 6V67 03.53 Redcar–Margam service, with a loaded coke train. Seen here on 16 May 2014.

EWS-liveried No. 66007 drags DB Schenker-liveried General Motors Type 5 Class 59 No. 59204 as it powers away from Eastleigh Yard while working the 6V41 14.47 Eastleigh–Westbury departmental service. Seen here on 13 May 2014.

No. 66199 passes through Whittaker Junction while in charge of the 6M03 Robeston–Bedworth Murco sidings loaded tank train. In the distance, the roof of GB Railfreight's No. 66720 can be seen after just arriving with a service from Felixstowe. Seen here on Friday 16 May 2014.

No. 66035 powers away from a signal check at Clitheroe while working the 18.55 6G35 Clitheroe Castle Cement–Bescot Down Side loaded cement tanks. The train is seen passing Taylors House at Clitheroe in the evening of 22 July 2014. This train is booked to run three days a week; Tuesdays, Thursdays and Saturdays.

No. 66206 climbs towards Melton Ross level crossings while in charge of the, slightly late running, 6F11 Immingham PAD 1–Cottam power station loaded coal train. Pictured here on 12 March 2014.

The pioneer EWS/DB Schenker Class 66 No. 001 powers past Burton upon Trent while working the 6D44 Bescot–Toton departmental service. The locomotive, along with sister No. 66002, are different from the other 248 EWS members of the class as they do not have buckeye coupling equipment, which restricts these locomotives to run on coal trains only. Seen here on 11 March 2014.

After the failure of Class 56 No. 56069 at Blackburn the previous night while in charge of the 6E73 Clitheroe Castle Cement–Healey Mills empty coal train, Class 66 No. 66001 was sent to rescue the train the following morning, which worked as the 0E73 from Warrington Arpley to Blackburn. The ensemble then departed from Blackburn goods loop around 500 minutes late. Seen here on 28 December 2003.

No. 66043 is seen shunting its train into the yard at Didcot while arriving with the 6A48 14.16 Bicester COD–Didcot TC MOD train. Seen here on 24 July 2013.

No. 66108 powers through the Scottish countryside while in charge of the 6Y15 Mossend Down Yard–Fort William enterprise train. The location is Bridge of Orchy, in Argyll and Bute, Scotland, and was pictured here on 5 June 2006.

No. 66149 powers down the slow lines at New Barnetby while running early with the 6F11 Immingham–Cottam power station loaded coal train. Seen here on 12 March 2014.

The first Class 66 to be repainted with the new DB Schenker red livery was No. 66152, which is seen here arriving at Doncaster station. The working was the 4L08 12.35 Wakefield Europort–London Gateway. Seen here on 18 March 2009.

Now exported to France for use by Euro Cargo Rail is No. 66244, and it is seen here passing Barnetby with the 4C72 Scunthorpe–Immingham MGR train. Photograph taken on 29 December 2003.

West Highland stag-stickered No. 66111 powers past Gisburn on the Ribble Valley line while working the 6L49 to Hellifield–Crewe Basford Hall rail drop train. Seen here on 14 March 2010.

No. 66227 sits in Clitheroe Castle cement works after working the 6Z84 from Redcar. The locomotive has now been exported to France and is seen here on 14 April 2004.

No. 66116 powers past Clay Mills Junction in Derbyshire while working the 6V66 Scunthorpe–Margam loaded steel slabs train. Seen here on 2 June 2006.

No. 66188 is seen sat on the Settle–Carlisle line at the north end of Blea Moor tunnel. The working had arrived earlier in the day from Carlisle and is seen here on 16 March 2006.

No. 66144 powers across the Bentham line while working the 6L64 Carnforth North Junction–Crewe Basford Hall engineering train via Settle Junction. Seen here on 14 March 2010.

DB Schenker-branded No. 66039 is seen shunting in Toton Yard while GBRF Class 66/7 66759 sits in the background awaiting its next duty. Seen here on 18 April 2015.

A very dirty No. 66053 is seen powering past the rarely used Whitley Bridge station while working the 6H43 Immingham–Drax loaded coal train. Seen here on 3 March 2015.

No. 66004 thrashes past Whitley Bridge while working the 6H51 Redcar–Drax power station loaded coal train. In the distance Nos 37425 and 37688 can be seen. Photographed here on 3 March 2015.

No. 66182 passes Burton Salmon with the 6H08 Humber–Drax loaded coal train. Seen here on 3 March 2015.

DB Schenker No. 66094 powers past Burton Salmon while in charge of the 6H84 Hull Biomass–Drax loaded coal train. Seen here on the 3 March 2015.

No. 66172 *Paul Melleney* is approaching Sherburn-in-Elmet station with the 6H26 Butterwell opencast mine–Drax power station loaded coal train. Seen here on 3 March 2015.

No. 66098 slowly crawls down the centre road at Cardiff Central while in charge of the 4E66 08.55 Margam–Redcar empty coal train formed of HTA hoppers. Seen here on 16 April 2014.

No. 66006 is seen approaching Whitley Bridge with the 4R17 Drax power station–Immingham empty biomass train. Photographed here on 10 March 2015.

No. 66132 passes Gisburn with the 4M00 Mossend Down Yard–Clitheroe Castle Cement empty cement tanks. Seen here on 18 March 2015.

Passing through the north Yorkshire countryside at Cononley is No. 66009 with the 6E73 Hunterston–Ferrybridge loaded coal train. Seen here on 7 April 2015.

No. 66150 arrives at Eastleigh station with the 6B38 Marchwood MOD–Eastleigh East Yard working. English Electric Bo-Bo Electro-Diesel Class 73 No. 73119 is seen sitting in the yard, photographed here on 13 May 2014.

Seen before the beautiful back drop of the Cumbrian landscape is EWS No. 66093 as it heads south at Docker on the West Coast Main Line with the 6K05 Carlisle Yard–Crewe Basford Hall departmental service. Seen here running early on 2 May 2008.

No. 66149 powers past Slindon in Staffordshire with the 6M48 Southampton Eastern Docks–Halewood car train. Seen here on 16 May 2014.

No. 66002 passing through Lostwithiel with the 6M72 St Blazey–Cliffe Vale enterprise train. Seen here on 27 June 2007.

DB Schenker No. 66093 powers past Burton Lane farmers crossing while in charge of the 8D06 Gascoigne Wood–Drax power station working. Seen here on 21 July 2014.

Equipped with additional lighting and drawgear for Lickey Incline banking duties is No. 66059 (together with classmates Nos 66055–058) and it is seen here passing Burton Lane farmers crossing while in charge of the 6H82 Immingham–Drax power station loaded biomass train on 21 July 2014.

No. 66122 passes through Hatton while working the 4M52 11.32 Southampton–Castle Bromwich car train. Seen here on 2 April 2014.

No. 66025 speeds through on the Up fast line at Doncaster station while in charge of the 4N07 Cottam power station–North Blyth empty HTA coal train. Photographed here on 21 August 2014.

No. 66105 slowly passes Long Preston while working the 4M00 Mossend Down Yard–Clitheroe Castle Cement empty cement tanks. Seen here on 5 June 2013.

No. 66141 powers past Whittaker Junction with the 6M03 Robeston–Bedworth Murco sidings loaded tank train. Photographed here on 7 September 2006.

The final DB Schenker Class 66 was No. 66250, photographed here passing Portway near Elford while in charge of the 6V97 Beeston–Cardiff Tidal loaded scrap train. Seen here on 7 September 2006.

Direct Rail Services

Direct Rail Services Class 66/3 No. 66303 passes Red Bank near Newton-le-Willows while in charge of the 6K05 Carlisle North Yard–Crewe Basford Hall departmental service. Seen here on 18 September 2014.

Ex-Fastline Rail No. 66301 is seen working for its new owners DRS, sitting at Blackburn station while working the 3J11 Carlisle Kingmoor–Carlisle Kingmoor RHTT train. The working goes via the Settle–Carlisle line, WCML, Cumbrian Coast and the Bentham line. The train is seen here on the evening of 10 November 2014.

Class 66/4 No. 66431 is seen restarting its train at Leyland after dropping off two colleagues who were route learning. The train is the 6K05 Carlisle North Yard–Crewe Basford Hall departmental service. Seen here on 8 August 2013.

Class 66/3 No. 66303 passes Barrow foot crossing near Whalley while working the 6K05 Carlisle North Yard–Crewe Basford Hall departmental service. Seen here on 12 September 2013.

Now run by Freightliner Poland, after being exported by its new owners (Freightliner), is former DRS No. 66412. The GM Type 5 is seen powering down the West Coast Main Line at Brock while in charge of the 4Z34 Coatbridge–Daventry container service. Seen here on 16 June 2008.

After winning the contract from Freightliner, Direct Rail Services GM Class 66 No. 66428 powers past Slindon in Staffordshire while in charge of the 6U77 Mountsorrel sidings to Crewe Basford Hall loaded ballast train. Seen here on a very sunny 12 June 2014.

The West Coast Main Line around Carnforth sees very little freight traffic in daylight hours due to a shortage of paths. However, No. 66429 is seen powering past the sea at Hest Bank while in charge of the 6K27 Carlisle North Yard–Crewe Basford Hall departmental service. Seen here on 21 May 2014.

Now run by DRS is ex-Fastline Rail Class 66/3 No. 66302, seen here passing Warrington Bank Quay with the 4S44 Daventry–Coatbridge container train. Seen here on 1 May 2014.

Stobart Rail-liveried No. 66414 *James The Engine* is seen powering past Hest Bank while in charge of the 4M16 Grangemouth–Daventry intermodal service. Seen here on 22 May 2010. The locomotive is now run by Freightliner and sports the new PowerHaul livery.

DRS shows off its latest revised livery as No. 66421 passes Taylors House at Clitheroe with the 0Z36 Carlisle Kingmoor sidings to Preston, via the Settle–Carlisle line, route learning trip. Seen here on 27 February 2015.

No. 66303 powers past the wooden bridge at Clitheroe while in charge of the weekday running 6K05 Carlisle North Yard–Crewe Basford Hall departmental service. Seen here on 12 September 2013.

Direct Rail Services No. 66304 passes through Clitheroe while working the 6K05 Carlisle North Yard–Crewe Basford Hall departmental service. In the distance Clitheroe Castle can be seen. Seen here on 17 July 2014.

No. 66431 heads down the long drag at Settle while working the 6K05 Carlisle North Yard–Crewe Basford Hall departmental service. The train runs from Monday to Friday and is seen here on 18 August 2014.

Direct Rail Services Class 66/4 No. 66407 is seen on the north side of Blea Moor tunnel with an engineering train. The train had worked from Carlisle North Yard earlier in the day. The date was 16 March 2006. The Class 66 has now moved to GBRF and is now numbered 66743.

Direct Rail Services No. 66421 powers the 6U77 Mountsorrel sidings–Crewe Basford Hall loaded stone train past Barrow on Trent. The train is pictured here on 11 March 2014.

No. 66434 passes Barton with the 6K05 Carlisle Yard–Crewe Basford Hall departmental train. Seen here at speed on the West Coast Mainline after being diverted due to engineering work on 10 October 2013.

Fastline

Seen here after being on long-term hire to Fastline from Direct Rail Services is No. 66434, still sporting its Fastline livery. The working is the 4M34 Coatbridge–Daventry container service, which is a DRS train. Photographed on 6 October 2010.

No. 66434 is seen powering past Tebay with the 6C37 Chirk Kronospan–Carlisle Yard empty log train. Seen here on 9 July 2010.

Freightliner

Freightliner No. 66548 slows down on the approach to Hellifield in preparation to cross onto the Ribble Valley Line while in charge of the 6M11 Hunterston High Level FHH–Fiddlers Ferry power station loaded coal train. Seen here on 23 July 2014. The location is Long Preston.

No. 66597 Viridor crosses over the pointwork at Fairwood Junction while being in charge of the 6A72 Fairwater Yard (Taunton)–Westbury HOBC train. Seen here on 14 May 2014.

Freightliner intermodal No. 66541 powers past Kings Sutton with the 4M62 Southampton Maritime–Hams Hall loaded container train. Seen here on 15 May 2014.

Intermodal Class 66/5 No. 66538 passes DB Schenker super tug No. 60019 at Kingsbury while in charge of the 4O54 Leeds FLT–Southampton loaded container service. Pictured here on 16 May 2014.

Freightliner No. 66522 pulls into Hellifield goods loop with the 6M21 Hunterston–Ratcliffe loaded coal train. Seen here on 29 April 2014. Classmate No. 66598 was dead in tow behind the leading locomotive.

Intermodal No. 66503 *The Railway Magazine* passes Sherburn-in-Elmet with the 4D07 Wilton FLT to Leeds FLT container train. Seen here on 8 April 2014.

Freightliner Heavy Haul Class 66/5 No. 66555 powers past New Barnetby with the 6M29 Immingham–Ratcliffe power station loaded coal. Seen here on 12 March 2014.

Freightliner Class 66/6 No. 66616 powers past Colton Junction heading towards Church Fenton with the 6Z66 Thrislington Steetley–Hunslet Tilcon loaded stone train. Seen here on 20 April 2013.

No. 66614 passes Hellifield pond with the 6Z68 Killoch Colliery–Drax power station loaded coal train. Seen here on 19 April 2008.

Freightliner Green Team No. 66565 is seen passing the now electrified Manchester Victoria while working the 6J44 Brindle Heath–Dean Lane binliner refuse train. Seen here on 2 May 2014.

Freightliner Intermodal No. 66542 passes through Eastleigh while in charge of the 4O54 Leeds FLT–Southampton loaded container service. Photographed here on 13 May 2014.

Looking rather good in the new Freightliner Powerhaul livery is Intermodal Class 66/5 No. 66504, seen here shunting at Millbrook station near Southampton. The locomotive would later work the 12.55 4M61 Millbrook FLT–Trafford Park container service. This was the first Class 66 to be painted in the outstanding new livery. Photographed here on 13 May 2014.

Coal trains around north and west Yorkshire are very busy due to the high demand of the three main power stations in Yorkshire, which are Ferrybridge, Eggborough and Drax. Freightliner No. 66596 passes through Burton Salmon in north Yorkshire while passing with the 6B53 Redcar–Eggborough loaded coal train. Seen here on 13 March 2014.

Even though the locomotive belongs to the same company, Freightliner Class 66/5 No. 66589 is normally seen on the Intermodal side of the company. However, a shortage of locos on the Heavy Haul side saw the rare appearance of the locomotive as it passes over the Bentham line in north Yorkshire. The working was the 6Y55 Horrocksford Junction at Clitheroe–Crewe Basford Hall engineering train via Carnforth. Photographed here on 23 November 2014.

Ipswich in Suffolk is known for being notoriously busy for container trains due to the nearby port of Felixstowe. Three of Freightliner's Class 66s, numbers 66540, 541 and 542, all sit in Ipswich Station Yard awaiting next duties and fuelling. The photograph was taken on 10 June 2008.

No. 66592 *Johnson Stevens Agencies* is seen powering past Slindon near Millmeece in Staffordshire. The working was the 4M61 Southampton MCT–Trafford Park FLT container service. Seen here on 12 June 2014.

A very unusual working was captured on 16 May 2014 when No. 66564 worked the 6Y19 Rugby–Tyne SS HOBC move. The train was routed via Coventry, Water Orton, Nuneaton, Derby and York.

No. 66594 motors through Shrivenham in Oxfordshire with the 09.58 4O51 Wentloog–Southampton container train. The train is seen here on 11 June 2014. The locomotive is named *NYK Spirit of Kyoto*.

Ex-Direct Rail Services (DRS) Stobart Rail No. 66414, which was named *James the Engine*, is seen working for its new owners Freightliner. The loco has now been painted into the new Powerhaul scheme. The train seen here, passing through Shrivenham near Swindon, is the 11.00 4L32 Bristol Freightliner Terminal–Tilbury container working. Photographed on 11 June 2014.

Re-geared Class 66/6 No. 66610 powers past Burton Salmon on the Castleton line while working the 6Z54 Oxwellmains Lafarge–Leeds Hunslet Tilcon loaded cement train. Seen here on 11 April 2014.

A busy location for the GM Type 5s is Doncaster. Here we have Freightliner Class 66/5 No. 66598 passing through Doncaster station on the fast lines while working the 6Y58 Applehurst Junction–Doncaster Belmont Yard engineering train. Photographed here on 15 August 2013.

No. 66546 powers past Docker on the West Coast Main Line while in charge of the 6C16 Crewe Basford Hall–Carlisle Yard departmental service. Photographed here on 2 May 2008.

Freightliner Heavy Haul No. 66552 *Maltby Raider* is seen after crossing the twenty-four arch Ribblehead Viaduct while working the 6E31 Killoch Colliery–Leeds Hunslet loaded coal train. Driver Chris Horner was in charge and is seen here on 24 January 2008.

No. 66597 passes through Doncaster station with the 6E45 Earles sidings–Drax power station loaded flyash. The train is seen here on 3 April 2015.

No. 66531 powers past Burton Salmon with the 6M49 Hull Kingston Terminal–Rugeley B power station, and is seen here on 13 March 2014.

Class 66/5 No. 66547 is looking a little scruffy as it passes Whitley Bridge with the 6R08 Immingham–Eggborough loaded coal train. Pictured here on 20 April 2013.

Seen sitting on the site of the former Gisburn railway station is the now-exported No. 66586, which has been sent to work for Freightliner Poland. The train had worked from Crewe Basford Hall earlier in the day and was the 6L58 to Horrocksford Junction at Clitheroe. Photographed here on 19 June 2008.

The first of the seven Freightliner Class 66/9s is No. 66951 and is seen here sitting in Crewe works during the open day back in September 2005. The Freightliner locomotive arrived into the UK in October 2004. Sister locomotive No. 66952 arrived in April 2004, and both are the first low-emission variant of Class 66.

Class 66/6 No. 66618 is seen sitting in the snow at Ribblehead while working the 6L58 Carlisle North Yard–Blea Moor engineering train. The locomotive is named *Railways Illustrated Annual Photographic Awards Alan Barnes*. The train was pictured here on 13 March 2006.

No. 66620 is seen approaching Long Preston after departing from the work site at Settle Junction with the return working to Crewe Basford Hall ballast drop train. Seen here on 24 March 2007.

Former DRS Class 66/4 No. 66419 is seen passing Whitley Bridge with the very late running 4R16 Drax power station–Immingham Dock CT. The locomotive is now run by Freightliner and is seen here on 10 March 2015.

No. 66557 is seen powering past Whitley Bridge while working the very short leg to Kellingley Colliery. The train is the 4A69 from Drax power station and was photographed here on 10 March 2015.

No. 66526 powers past Whitley Bridge with the 6E94 23.53 Hunterston High Level FHH–Drax AES (FLHH) loaded coal train. Seen here on 9 March 2015. The locomotive is named after Driver Steve Dunn, who sadly lost his life in the Great Heck smash involving No. 66521.

Freightliner No. 66610 passes Chapel-en-le-Frith with the 12.08 6M03 Barrow Hill sidings FHH–Tunstead empty stone train. The train is seen here on 5 June 2008.

No. 66553 passes Knottingley while in charge of the 6E94 23.53 Hunterston High Level FHH–Drax AES (FLHH) loaded coal hoppers. The train is seen here on 11 April 2014.

No. 66552 *Maltby Raider* heads the 6Y13 Immingham Dock to Ferrybridge power station loaded coal train, seen here passing Burton Salmon on 11 April 2014.

No. 66532 is approaching Reading with the 4L32 11.00 Bristol Freightliner Terminal to Tilbury container train. The loco was photographed while travelling on the 1C83 London Paddington–Plymouth service. Seen here on 22 July 2013.

The East Lancashire Railway at Bury held their diesel gala over the weekend of 10/11 of July in 2004. The main attractions were Nos 66558 and 66565, and they are seen here on the Saturday as they power past Burrs Country Park near Bury with a service bound for Rawtenstall.

Freightliner No. 66613 powers past Flag Lane at Preston with the 6Y15 Preston Fylde Junction–Crewe Bamford Hall Engineering train. Class 66/5 No. 66526 was on the rear, out of sight in this photo. Seen here on 18 March 2015.

No. 66528 powers past Taylors House at Clitheroe while in charge of the 05.49 6M11 Hunterston High Level FHH–Fiddlers Ferry power station loaded coal train. The train was photographed here 19 August 2014. The Class 66 has since been repainted into Freightliner Powerhaul colours.

The first of Freightliner's 118-strong fleet of Class 66s to be repainted into the new colour scheme of Powerhaul was No. 66504, and is seen here approaching Eastleigh with the 12.55 4M61 Southampton Millbrook–Trafford Park FLT container service. Seen here on 13 May 2014.

Sporting a distinctive and unique livery is No. 66522 as it sits in Hellifield goods loop while in charge of the 6M21 Hunterston–Ratcliffe loaded coal train. Sister locomotive No. 66598 was dead in tow and is seen here on 29 April 2014.

No. 66534 powers through Rugby station with the 4M88 Felixstowe North–Crewe Basford Hall container train. Seen here on 24 July 2009.

Bardon Aggregates-liveried No. 66623 *Bill Bolsover* is seen arriving into Crewe Basford Hall yard with the 13.42 6U77 from Mountsorrel sidings. The train is now operated by DRS and is seen here on 6 April 2009.

Freightliner partnered other railway companies in providing traction for Pathfinder Tours. Here we see No. 66608 at Morecambe station after working down the Heysham branch. The railtour was the 1Z37 05.40 Reading to Preston via Clitheroe and Morecambe. No. 66951 was on the front of the train and was pictured here on 30 May 2005. The tour name was 'The Multi-Coloured Swap Shop' and also included Nos 59206, 60088 and 92010.

Ex-DRS Malcolm-liveried Class 66/4 No. 66412 is seen now working for Freightliner as it passes Shrivenham with the 4V50 Southampton–Cardiff Wentloog container train. Photographed here on 18 October 2011. The locomotive has now been exported by Freightliner for use in Poland.

The first Freightliner Class 66 is No. 66501, seen here passing Kings Sutton while in charge of the 4M62 Southampton Maritime to Hams Hall loaded container train. Photographed here on 17 June 2010.

No. 66598 powers past Hest Bank in the powerful summer sun with the 6Z16 Carlisle Yard–Crewe Basford Hall departmental service. Seen here running very early on 10 July 2014.

GBRF

GBRF 'ghost train' No. 66703 passes Whitley Bridge while in charge of the 6H36 Tyne Dock–Drax loaded coal train. Photographed here in the early hours of 13 March 2014.

GB Railfreight No. 66739 powers past Slindon with the 6K50 Toton–Crewe Basford Hall departmental service. Photographed here on 12 June 2014.

No. 66720 passes Slindon with the 4F01 Ironbridge power station–Seaforth containerised biomass train. Seen here on 12 June 2014.

GB Railfreight Class 66/7 No. 66716 powers up the gradient at Burton Salmon with the 6H30 Tyne Dock–Drax loaded biomass train. Photographed here on 11 April 2014. The engine is named *Locomotive and Carriage Institution Centenary 1911–2011*.

No. 66750, which is now run by GB Railfreight, is seen powering past Whitley Bridge with the 6C71 Hull Coal Terminal GBRF–Eggborough power station loaded coal train. Photographed here on 13 March 2014. This locomotive was numbered 6606 and was new to ERS Railways in the Netherlands in 2003, who used it up until 2008 when it transferred to Freightliner Poland, followed by Crossrail Benelux, after which it was due to go to Swedish firm Rush Rail and was painted in its blue livery, which it retains. The deal later got cancelled, meaning the locomotive was transferred to GBRF in June 2013.

After creating the London Underground map in 1933, designed by Harry Beck, former GB Railfreight Class 66 No. 66721 sports a very special London Underground livery with graphic images as effect. The locomotive is seen powering away from Knottingley with the 4D61 Ferrybridge power station–Doncaster Down Decoy empty coal train. Seen here on 13 March 2014. The loco is named *Harry Beck*.

After carrying the Metronet livery since being delivered, GBRF No. 66718 now sports a new London theme with graphics showing the various modes of London transport. The locomotive is named after Sir Peter Hendy OBE, and is seen departing from Eggborough power station while working the 4N57 to Tyne Coal Terminal. Seen here on 10 March 2015.

After arriving into the country in 2005, ex-Freightliner Class 66 No. 66580, now known as No. 66740 and carrying the new GBRF livery, is seen passing Colton Junction just south of York with the 4N47 Drax AES GBRF–Tyne Coal Terminal empty coal train. Seen here on 20 April 2013.

First-liveried No. 66723, now named *Chinook*, motors down the Great Eastern Main Line at Brantham while working the 10.46 4M23 Felixstowe North GBRF–Hams Hall GBRF container train. Seen here on 10 June 2008.

General Motors Class 66 No. 66744 is seen working for GB Railfreight after having worked for DRS, Advenza and Colas. The locomotive has been numbered three times, firstly renumbered from 66408 to 66843 and now holds its third number of 66744. The train is pictured here at Colton Junction while working the 07.32 4N36 Drax AES–Tyne Coal Terminal on 26 May 2012.

GBRF No. 66730 slows for a signal as it approaches Whitley Bridge railway station with the 6H93 Tyne Dock Coal Terminal–Drax AES loaded coal train. Seen here on 26 May 2012.

A mixture of locomotives can be seen here working as a trio as Nos 66725, 66755 and 66751 motor past Tallington Junction on the East Coast Main Line with the 6H78 Doncaster Down Decoy–Peterborough Virtual Quarry movement. On the rear of the train, two more GBRF Class 66s, Nos 66743 and 731, can just about be seen. The train was captured here on 27 September 2014.

No. 66716 arrives at Doncaster after working the 5X89 11.30 Slade Green–Doncaster Wabtec unit drag. The unit was No. 465902 and is seen here on 27 September 2014.

No. 66709 powers past Garsdale in its old Medite livery while working the 4E13 Newbiggin British Gypsum–Doncaster Down Decoy empty gypsum train. Seen here on 15 July 2006. The locomotive has since been repainted, carries the name *Sorrento* and has a picture of a MSC boat on the side.

After only arriving in the country in late 2014, here we can see one of the latest additions to GBRF's strong fleet of locomotives as No. 66758 passes through Doncaster railway station. The working was the 4D79 Selby Potters–Doncaster Down Decoy GBRF wagon move. The train then went on to Trafford Park later in the day. Seen here on 7 February 2015.

The now electrified line between Manchester and Liverpool means the area this photo was captured (near Earlestown) has gone due to overhead live wires. Here, we see No. 66732 passing through Earlestown railway station while in charge of the 13.00 4F61 Ironbridge power station–Tuebrook sidings empty biomass train. Seen here on 18 September 2014.

Former Crossrail AG railway company Class 66 No. 66749 passes Sherburn-in-Elmet while working for its new company GB Railfreight. The train was the 11.51 6Y89 North Blyth–Ferrybridge power station loaded coal train. Seen here on 13 March 2014.

The first of the GB Railfreight fleet was No. 66701, and it is seen here still in the original GB Railfreight livery. The locomotive was working the 0D16 Immingham Mineral Quay–Doncaster Down Decoy. Seen here on 26 May 2012; the location was New Barnetby.

No. 66713 *Forest City* is seen powering past the site of the former Sudforth Lane signal box which, unfortunately, was demolished in the spring of 2014. The working was the 4N94 17.10 Eggborough power station–Tyne Coal Terminal. Seen here in the late afternoon on the 12 March 2014.

GBRF First-liveried Class 66/7 No. 66725 *Sunderland* is seen crossing the points to head down Platform 1 at Doncaster railway station while in charge of the 4L78 12.00 Selby–Felixstowe North container train. Seen here on 12 March 2015.

No. 66739 *Bluebell Railway* is seen passing Whitley Bridge with the 4N30 15.29 Drax AES–Tyne Coal Terminal empty biomass train. Seen here on 10 March 2015.

No. 66720 is seen wearing a very bizarre livery called 'Night and day', designed by a seven-year-old girl after she won a competition. The locomotive is seen adding colour to the open countryside at Burton Salmon while working the 6H30 Tyne Coal Terminal–Drax loaded biomass train. Seen here on 21 July 2014.

Former Direct Rail Services Class 66/4 No. 66409 is now seen working for GBRF and has been renumbered 66745. The train was the 0D63 Tyne Dock–Doncaster Roberts Road light engine move. This train ran due to teammate No. 66717 suffering engine problems. The location is Sherburn-in-Elmet and this image was captured on 7 August 2014.

No. 66730 powers past Slindon in Staffordshire with the weekday 6K50 Toton–Crewe Basford Hall departmental service. Seen here in the late afternoon on 16 May 2014.

First-liveried No. 66724 is seen slowly heading through the busy city of Manchester as it approaches Manchester Piccadilly station. The train was the 14.18 Trafford Park Euro Terminal–Felixstowe North. Seen here on 7 May 2014. In the distance, Freightliner PowerHaul No. 66504 can be seen.

GB Railfreight No. 66716 powers away from a signal check at Knottingley while working the late-running 15.29 4N30 Drax AES–Tyne Coal Terminal empty biomass train. Seen here on 11 April 2014. DB Schenker Class 09 shunter No. 09201 is seen on the right after working the 8K77 Milford Junction to Knottingley wagon move earlier in the day.

Ex-Netherlands-based No. 66749 is now operated in the United Kingdom by GB Railfreight and is seen here passing through Knottingley with the 06.25 6H12 Tyne Coal Terminal–Drax AES loaded biomass train. Seen here on 11 April 2014.

GBRF First-liveried Class 66/7 No. 66727 passes rival company Freightliner's No. 66598 while working the 11.28 4N80 Doncaster Down Decoy–Tyne Coal Terminal empty coal train. Seen here on 15 August 2013.

GB Railfreight Class 66s are in many different liveries and here we see the First Group livery, also known as the Barbie livery. No. 66725 passes through Sherburn-in-Elmet with the 12.49 6B64 Tyne Coal Terminal–West Burton power station loaded coal train. Seen here on 13 March 2014.

No. 66703 *Doncaster PSB 1981–2002* is seen approaching Whitley Bridge train station while working the 09.41 4D93 Doncaster Down Decoy–Drax AES gypsum train. Seen here on 9 March 2015.

No. 66743 screams past Whitley Bridge while in charge of the late running 05.00 6H99 Tyne Coal Terminal–Drax AES loaded coal train. Seen here on 9 March 2015.

No. 66762 slowly rolls down the Rylstone branch while working the 11.10 Rylstone Tilcon to Wellingborough Up Tc loaded stone train. The locomotive is less than twelve months old and is seen here on 18 March 2015.

No. 66723 *Chinook* powers past West Bank Hall level crossings on the Drax branch. The working is the 09.49 6H30 Tyne Coal Terminal–Drax AES and was pictured here on 7 August 2014.

The very bizarre looking No. 66720 powers past Ribblehead station while working the 12.35 4C77 Fiddlers Ferry power station to New Biggin loaded gypsum train. Seen here on 4 September 2012. The train is now run by DB Schenker and runs later in the evening.

Class 66/7 No. 66753 *EMD Roberts Road* is seen approaching Doncaster near journeys end while it works the 4D63 Tyne Coal Terminal–Doncaster Down Decoy loaded coal train. Photographed here on 5 April 2015.

No. 66756 powers past Clarborough Junction after just starting its journey from West Burton power station, the working was the 16.00 6V80 to Portbury Coal Terminal. The train is seen here on 5 April 2015.

No. 66717 *Good Old Boy* is seen passing Barnetby, running five minutes late with the 4R73 Cottam power station–Immingham HIT empty coal service. Seen here on 4 April 2015.

The race is on at Trimley on the Felixstowe branch when two GBRF Class 66s race for a path. No. 66724 is seen passing through Trimley station while working the 10.46 4M23 Felixstowe North–Hams Hall container service. Sister No. 66719 can be seen in the distance while working the 0Y23 Felixstowe North–Ipswich Reception. Seen here on 29 March 2007.

Class 66/7 No. 66740 powers through Cononley while working the 11.00 6M19 Rylstone Tilcon–Wellingborough Up Tc. Photographed here in the scenic Yorkshire countryside on 7 April 2015.

Metronet-liveried No. 66718 powers past Long Preston as it heads towards the 'Long Drag' on the Settle–Carlisle line. The working is the 12.35 4C77 Fiddlers Ferry power station–New Biggin loaded gypsum train. Seen here on 26 March 2013.

No. 66709 *Sorrento* powers past Eastleigh while in charge of the 4Y19 Mountfield sidings–Southampton W. Docks gypsum train. Seen here on 16 May 2015.

A uniquely liveried No. 66705 *Golden Jubilee* powers up the hill at Rylstone with the 6M19 1110 Rylstone Tilcon–Wellingborough Up Tc loaded stone wagons. Seen here on 3 February 2015.